ROSEN Verified
NATURAL DISASTERS

EARTHQUAKES

Jill Keppeler

ROSEN
PUBLISHING

Published in 2023 by The Rosen Publishing Group, Inc.
2544 Clinton Street, Buffalo, NY 14224

First Edition

Editor: Greg Roza
Book Design: Michael Flynn

Photo Credits: Cover Perfect Gui/Shutterstock.com; (series background) PinkPueblo/Shutterstock.com; p. 5 austinding/Shutterstock.com; p. 7 OSweetNature/Shutterstock.com; p. 9 Peter Hermes Furian/Shutterstock.com; p. 11 YJ.K/Shutterstock.com; p. 13 (top) VectorMine/Shutterstock.com; p. 13 (bottom) https://commons.wikimedia.org/wiki/File:Seismogram_on_historical_seismometer_at_Institute_of_Geophysis_of_the_Czech_Academy_of_Science_(14).jpg; p. 14 Andrey VP/Shutterstock.com; p. 15 dizain/Shutterstock.com; p. 17 Evgeny Haritonov/Shutterstock.com; p. 19 LeStudio/Shutterstock.com; p. 20 tunasalmon/Shutterstock.com; p. 21 https://en.wikipedia.org/wiki/File:EQs_1900-2013_worldseis.png; p. 23 courtesy of USGS; p. 25 Virrage Images/Shutterstock.com; p. 27 Frans Delian/Shutterstock.com; p. 29 https://commons.wikimedia.org/wiki/File:Port_Royal_earthquake_1692_by_Jan_Luyken_and_Pieter_van_der_Aa.jpg; p. 30 https://commons.wikimedia.org/wiki/File:CharlesRichter.jpg; p. 31 Feng Cheng/Shutterstock.com; p. 33 (top) Tupungato/Shutterstock.com; p. 33 https://commons.wikimedia.org/wiki/File:Valdivia_after_earthquake,_1960.jpg; p. 35 https://en.wikipedia.org/wiki/File:Post-and-Grant-Avenue-Look.jpg; p. 37 hedgehog94/Shutterstock.com; p. 39 Pixel-Shot/Shutterstock.com; p. 41 Microgen/Shutterstock.com; p. 43 SeventyFour/Shutterstock.com; p. 45 Mila Supinskaya Glashchenko/Shutterstock.com.

Cataloging-in-Publication Data

Names: Keppeler, Jill, author.
Title: Earthquakes / Jill Keppeler
Description: New York: Rosen Publishing, [2023] | Series: Rosen Verified : Natural Disasters | Includes index.
Identifiers: LCCN 2022028993 | ISBN 9781499469493 (library binding) | ISBN 9781499469486 (pbk) | 9781499469509 (DLC) 2022028994 (ebook)
Subjects: LCSH: Earthquakes - Juvenile literature
Classifications: LCC QE521.3 .K457 2023 | DDC 551.22

Manufactured in the United States of America

CPSIA Compliance Information: Batch #CWRYA23. For further information contact Rosen Publishing at 1-800-237-9932.

CONTENTS

WHEN THE EARTH SHAKES

They happen every day. They can happen anywhere. They can be natural or caused by human actions. They can be unnoticeable or cause huge amounts of harm and pain. They can even cause other natural disasters—sudden events in nature that cause much **damage**. They're a fact of life here on Earth. They're earthquakes!

These disasters are when the surface of the planet shakes. This is when there's a great release of energy, or power, inside Earth. Scientists can't prevent earthquakes. They also can't **predict** earthquakes yet. However, they're working to find ways to warn people when a bigger quake is on the way. They're also working to find ways to help people during quakes. Thanks to scientists, we can now create buildings, even very tall ones, that are safer. They bend rather than break when an earthquake hits.

Earthquakes can cost a lot of money in damage to property. They can also hurt or kill many people. Learning more about these natural disasters can help you understand them. It can also help you understand what to do if an earthquake happens where you are. Read on to find out more.

Earthquakes can make roads collapse, or break apart and fall.

UNDER THE EARTH

To understand earthquakes, you need to know about how Earth is built, deep under our feet. There are four main layers. These are the crust, the mantle, the outer core, and the inner core. The crust and the top part of the mantle make up a shell around Earth.

This shell is called the lithosphere. It's not one piece, though. It's made up of many pieces. These pieces are called tectonic plates. They're always moving very, very slowly across the surface of the planet.

AT THE CORES

The outer core of Earth is mostly the metals iron and nickel. The iron causes the planet's magnetic field. The inner core is mostly solid metal. It's very dense, or tightly packed together.

THE LAYERS OF EARTH

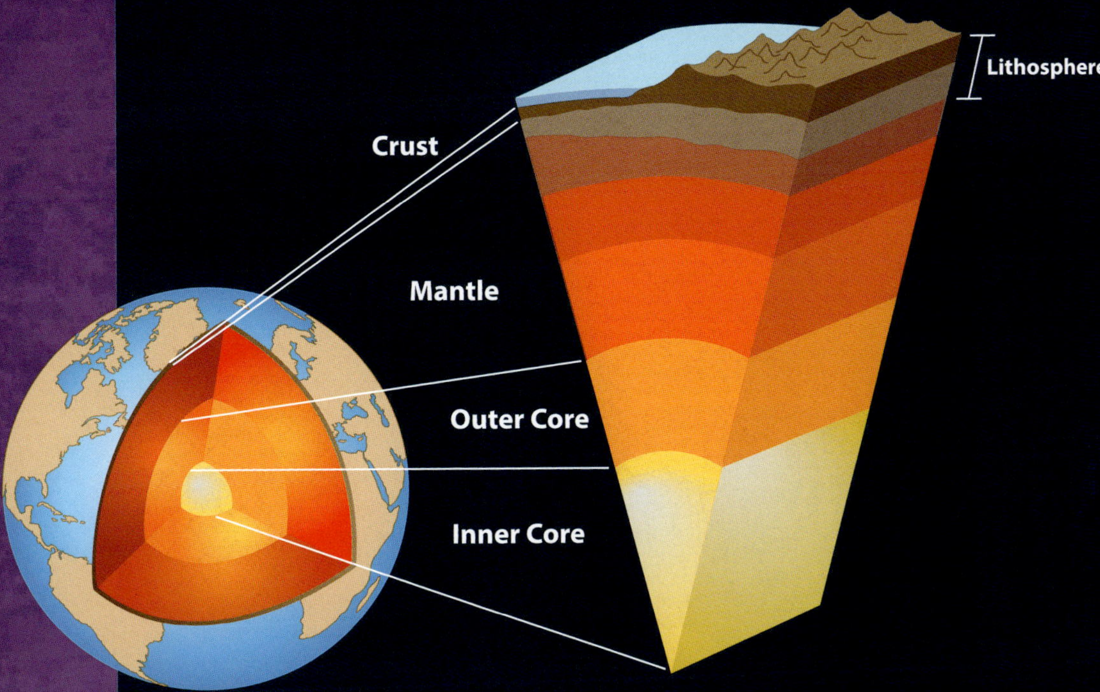

CRUST: A ROCKY LAYER OF EARTH.

LITHOSPHERE: A LAYER THAT INCLUDES THE CRUST AND THE VERY UPPER PART OF THE MANTLE.

MANTLE: A MOSTLY SOLID LAYER.

OUTER CORE: MOSTLY MOLTEN, OR MELTED LAYER.

INNER CORE: MOSTLY SOLID.

The lava that comes out of **volcanoes** is molten rock. It comes from the lower part of the crust or the upper part of the mantle. Underground, lava is called magma.

PIECES OF THE PUZZLE

Earth has many tectonic plates. They can be different thicknesses. Younger parts of the lithosphere under the ocean can be as little as 9.5 miles (15 km) thick. Very old pieces of the **continents** can be more than 124 miles (200 km) thick. The average is about 78 miles (125 km) thick.

There are seven main tectonic plates. They cover nearly 95 percent of Earth's surface. There are also eight smaller ones, plus many tiny plates.

ALFRED WEGENER

Alfred Wegener was a German scientist. He was born in 1880. He came up with the hypothesis, or idea, of continental drift. This means that the continents are slowly drifting across Earth. Few people believed him then. Today, his idea is the basis for plate tectonics.

EARTH'S TECTONIC PLATES

The biggest plate is the Pacific plate. It's under the Pacific Ocean. The North American plate is the second biggest.

TECTONIC PLATES OF EARTH

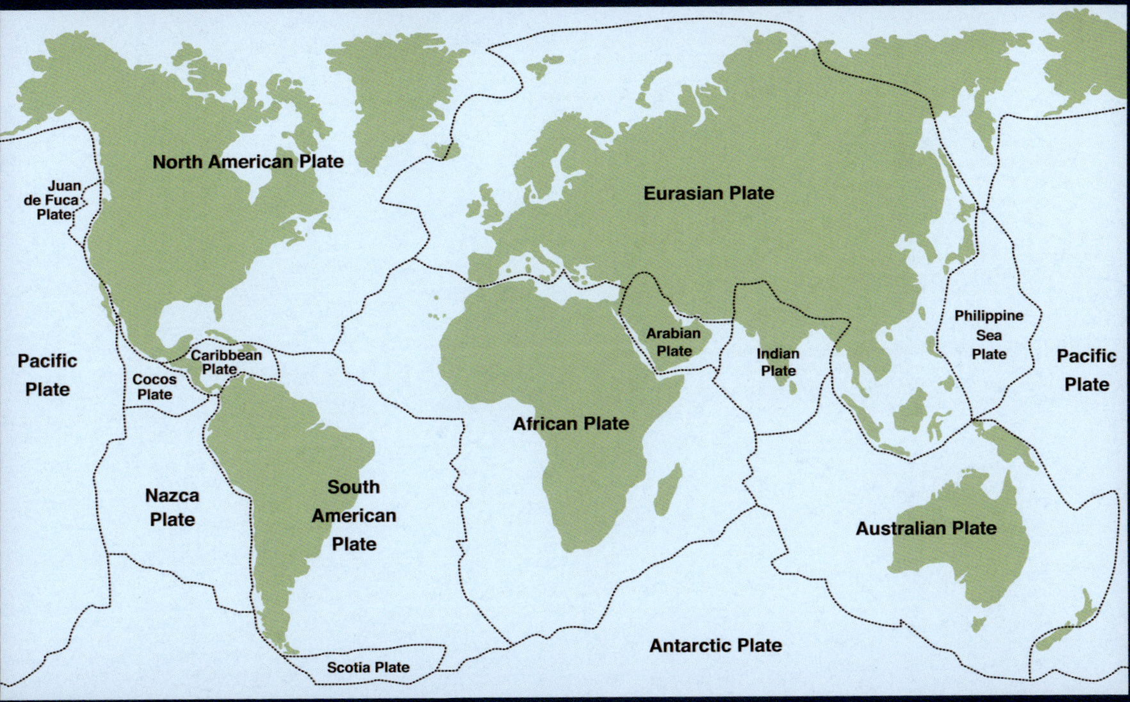

FLOATING ROCK?

Tectonic plates move because they float on the softer rock below them. They move very slowly: about 2 to 4 inches (5 to 10 cm) a year.

WHY DO QUAKES HAPPEN?

So, now you know what tectonic plates are. You know that they move across Earth. Earthquakes happen when the plates move around each other. They slide past each other. They bump into each other. Sometimes they slip under or over each other.

Sometimes the plate edges, or faults, get stuck together. The plates keep moving, though. The energy builds up. In time, the edges finally move. The energy is released in waves called seismic waves. They shake the earth.

MAN-MADE QUAKES

This book is about natural earthquakes. However, human actions can cause earthquakes too. These actions are usually caused by people getting rid of wastewater by putting it in the ground. They can be caused by ways people search for oil and gas too.

VERIFIED

The United States Geological Survey (USGS) is a science agency, or group. It's part of the Department of the Interior. This department oversees the country's lands and natural **resources**. The USGS has a lot of information about earthquakes on its website. Visit it at **usgs.gov**.

Energy often moves in waves. These look a little like the kind of waves you can see in water. Seismic waves move out from the site of the fault. It's a lot like dropping a stone into a pool.

RECORDING AND MEASURING

People use instruments called seismographs to record earthquakes. These recordings are called seismograms. Using seismograms, scientists can figure out how big an earthquake was. The size of the quake depends on the size of the fault. It also depends on the amount the edges move.

Scientists can also use seismograms to figure out where an earthquake took place. Quakes are recorded by a seismic network. This is a group of seismograms that are connected. They each take readings of the earthquake.

ON SOLID GROUND

Sometimes the earth shakes for other reasons. Maybe a big truck passes by. Maybe a new building is being built. That's why seismographs are usually kept in isolated, or very separate, spaces. That way, other things can't affect them. They're also often connected to bedrock. This is a very solid rock under the surface of the ground.

SEISMOGRAPH

A seismograph has a base that's on the ground and a hanging weight. When there's an earthquake, the base shakes. The weight doesn't. It records the difference.

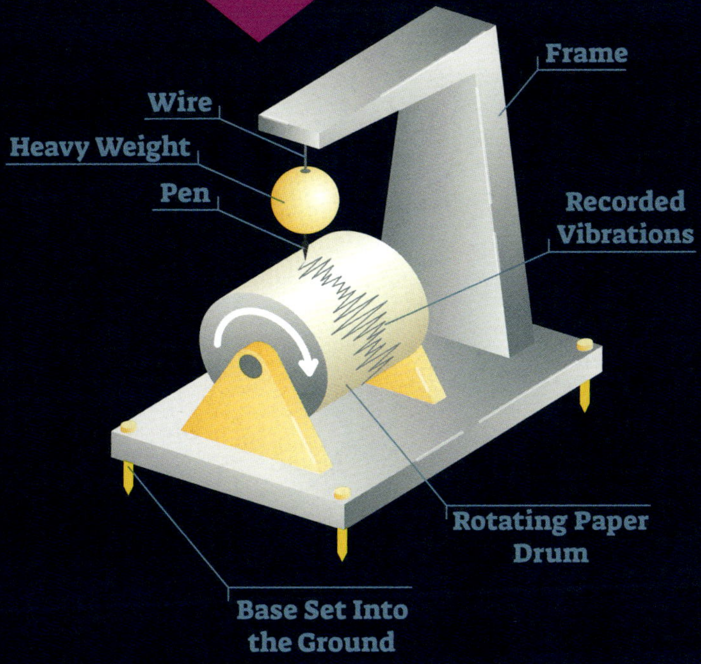

In a seismogram, the length of the moving line marks the size of the fault. The difference from a straight line marks the amount the plate edges moved.

HOW STRONG?

The size of an earthquake is called its magnitude. An earthquake's intensity is a measure of how hard the shaking feels. It's also a measure of the damage it causes. Magnitude is always the same for a given earthquake. Intensity is different depending on the location.

HOW MANY?

Dozens of earthquakes happen every day. Most of these, humans don't even feel. There are about 16 to 20 major earthquakes (with a 7 magnitude or above) each year.

There are different scales for measuring magnitude. Scientists used to use the Richter scale. Today, however, they usually use the moment magnitude scale. It gives a better estimate, or educated guess, of the size of an earthquake.

THE MOMENT MAGNITUDE SCALE

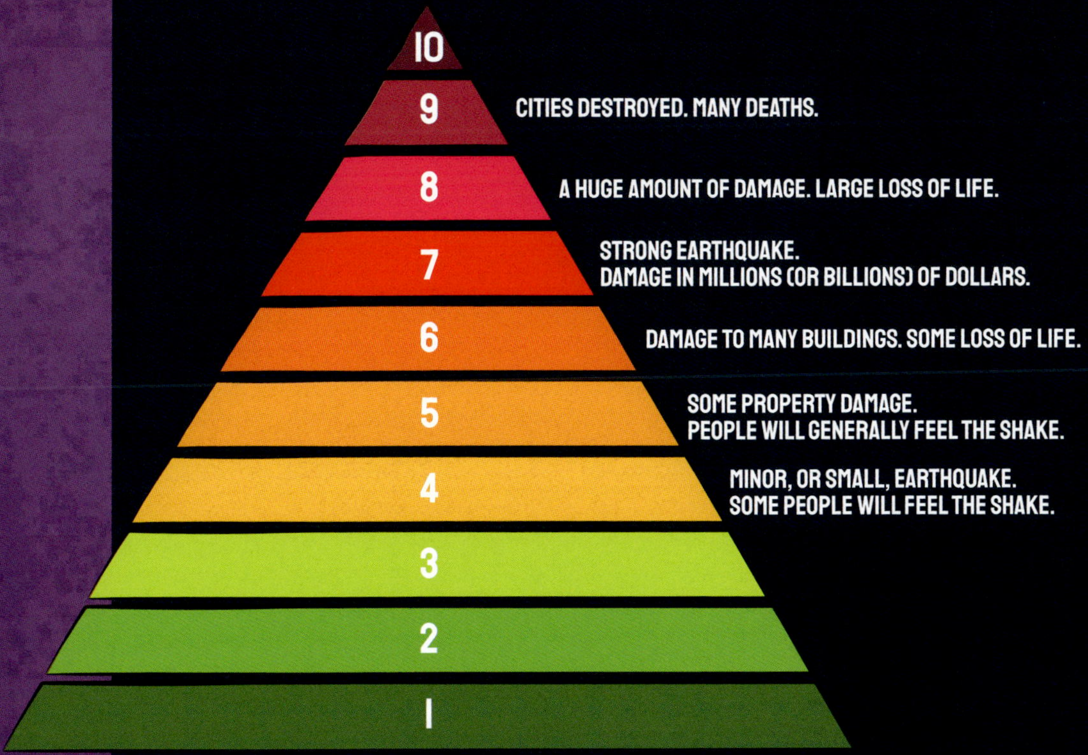

The bigger the number on the moment magnitude scale, the bigger the earthquake.

EARTHQUAKE SCIENTISTS

Scientists who study earthquakes are called seismologists. They also study related disasters such as volcanic **eruptions**. They apply what they learn to many things. This could include creating safer buildings. It can also include safely looking for resources underground.

Seismologists need to go to college. They often need an advanced **degree**. They need to think about things carefully and solve problems. They need to know about geology, math, and **physics**. Seismologists often also need to know about computers.

FAST FACT

Seismology is the science that deals with earthquakes. Seismologists study earthquakes. Seismographs record earthquakes. See what those words have in common? "Seismo" comes from the Greek word *seismos*. This word means, of course, "earthquake"!

VERIFIED
The U.S. Bureau of Labor Statistics is part of the U.S. Department of Labor. It provides numbers and other information about jobs in the United States. Its website offers pages with information about many different jobs. Visit it at **bls.gov**.

Seismology can be closely related to geology. This is the study of rocks and soil to learn more about Earth. Both kinds of scientists may spend time in the field. This means they're outside studying things in nature.

SEEING THE FUTURE?

Scientists can predict many natural disasters to some extent. But they can't predict strong earthquakes. They can sometimes figure out the probability, or chances, of one happening in a place over a number of years. A real prediction, though, needs the time, date, location, and strength.

Some people tell stories about how animals can predict earthquakes. They say animals head for safety a few days before a quake. This isn't **reliable**, though. Some people say they have a sense for earthquakes. This also isn't proven.

WEATHER PREDICTION

Some natural disasters can be somewhat predicted. These usually have to do with weather. Meteorologists are scientists who study weather. They can look at the air and temperature, or how hot or cold something is, to predict tornadoes and hurricanes. It's not a perfect science, however.

A tornado is a destructive storm. That means it causes great damage. Strong winds move around a middle point. This forms a funnel. Hurricanes are also destructive storms. They have very strong winds too. They form over the ocean.

VERIFIED

The National Weather Service (NWS) of the United States works to predict weather events. This includes dangerous ones such as tornadoes and floods. Learn more at **weather.gov**.

EARTHQUAKE CENTRAL

Earthquakes can happen anywhere. However, strong ones tend to happen in certain areas. Most larger earthquakes take place around the edges of the Pacific Ocean. This is also around the edges of the Pacific plate. This area is sometimes called the Ring of Fire.

RING OF FIRE

About 80 to 90 percent of Earth's largest earthquakes take place in the Ring of Fire. The region also has many volcanoes. This includes Mount Saint Helens in Washington State.

The second biggest area for earthquakes is the Alpide earthquake belt. This area goes from Java to Sumatra in Indonesia. It also goes through the Himalayan mountains and the Mediterranean area. It extends into the Atlantic Ocean.

GLOBAL EARTHQUAKES

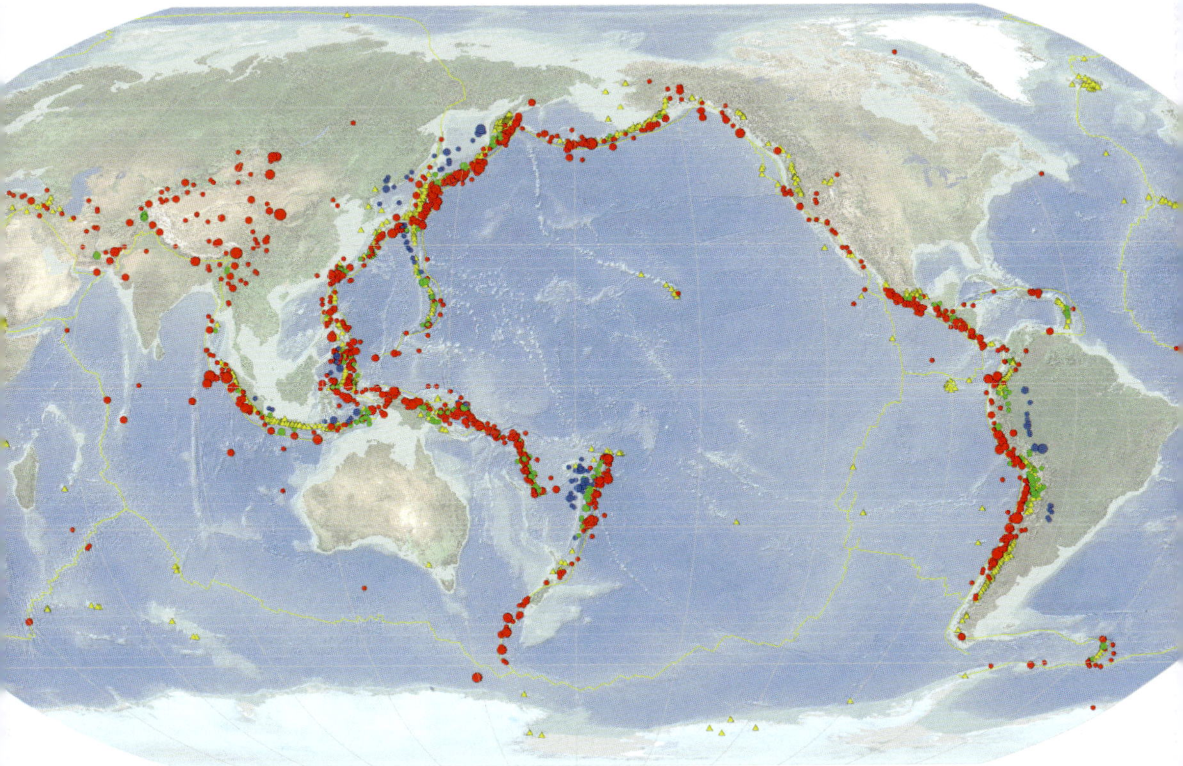

This map from the USGS shows where bigger earthquakes took place from 1900 to 2013. Compare it to the Ring of Fire map.

IN THE UNITED STATES

California has more earthquakes that cause damage than any other U.S. state. Alaska also has a great number of earthquakes. They're not the only states with bigger quakes. People in other western states of the country are at risk too. This includes Washington, Oregon, Nevada, and Utah.

However, other states also have many people in danger areas. This includes Tennessee, South Carolina, Arkansas, Missouri, and Illinois. About 28 million Americans live in areas with a high chance of damaging quakes.

FAST FACT

Florida and North Dakota are the states with the lowest risk of quakes. However, this doesn't mean they're impossible! Worldwide, Antarctica is the continent with the fewest quakes.

FAST FACT

The United States has 50 states and 14 **territories**. In all, nearly half of all Americans are at some risk from earthquakes.

UNITED STATES AND EARTHQUAKE RISK

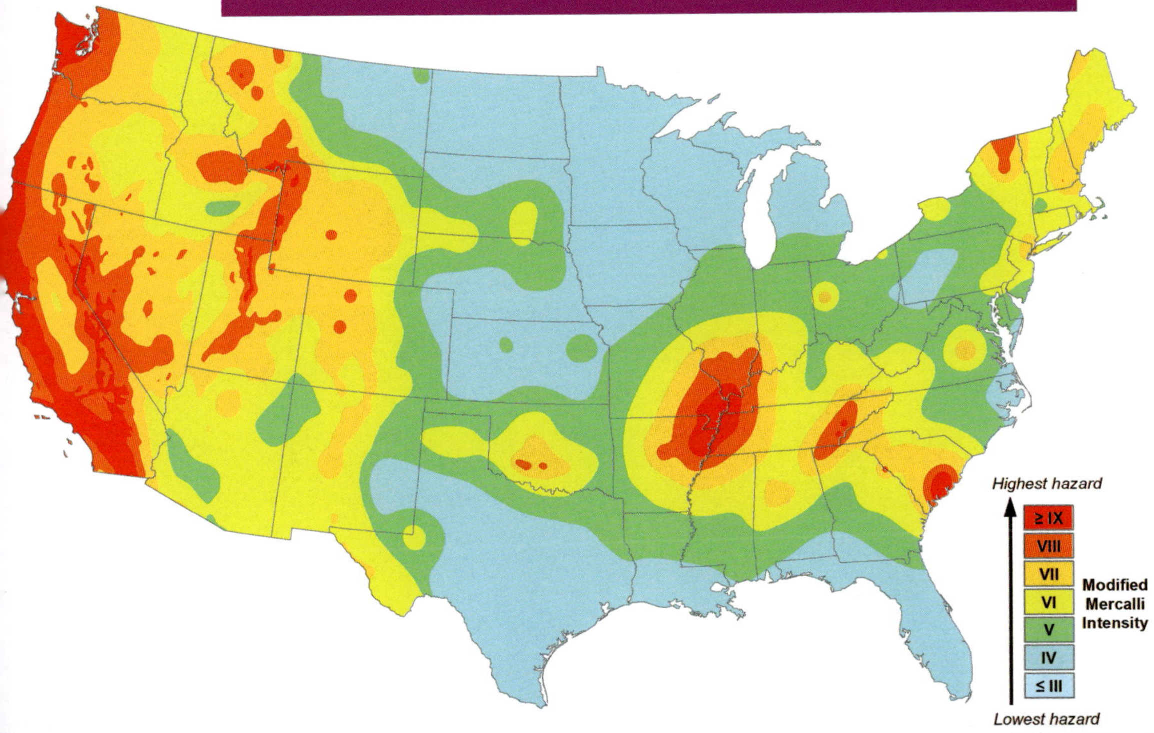

The western edge of the United States is part of the Ring of Fire region. There are some other earthquake areas too, however. New Madrid, Missouri, is in the middle of the red spot in the Midwest.

BEFORE AND AFTER

Sometimes an earthquake has smaller earthquakes that come before it. These are called foreshocks. They can't be used to predict bigger quakes, though. That's because scientists can't tell they're foreshocks until the bigger quake happens. This bigger quake is called the mainshock.

There are aftershocks after the main quake. These are smaller quakes in the same spot. They can go on for a long time. They can even continue for years! These happen as a fault **adjusts** after a mainshock.

WHAT'S A SWARM?

Quakes can also happen in swarms. Swarms are a cluster, or group, of smaller earthquakes. They happen around the same spot in a certain period of time. There's no mainshock, foreshocks, or aftershocks.

VERIFIED

Yellowstone National Park is part of the U.S. national park system. The National Park Service (NPS) runs these parks. To learn more about the NPS, visit **nps.gov**. For the Yellowstone National Park website, visit **nps.gov/yell**. And to learn more about Yellowstone earthquakes, visit **www.nps.gov/yell/learn/nature/earthquakes.htm**.

Yellowstone National Park sometimes has earthquake swarms. People can't feel most of them. They're centered around the Yellowstone Caldera, or volcanic crater. Part of Yellowstone Lake, seen here, is in the caldera.

A GREAT WAVE

Strong earthquakes can have many aftereffects. There may be aftershocks. There may be collapsed buildings or cracks in the ground. They may cause **landslides** or mudslides. But one of the deadliest possible aftereffects is a tsunami. These huge water waves are caused by undersea quakes or volcanoes.

A quake in the middle of the ocean may hurt no one directly. However, a tsunami can cause widespread destruction and death if it reaches shore. Most tsunamis happen in the Ring of Fire.

TSUNAMI NUMBERS

- Tsunami waves can be more than 100 feet (30.5 m) tall when they hit land.
- Tsunamis can move across the ocean at up to 500 miles (805 km) per hour.
- About 700 million people live in areas affected by tsunamis.
- Between 1998 and 2017, tsunamis caused more than 250,000 deaths.

THE 2004 INDIAN OCEAN TSUNAMI

The deadliest tsunami in known history happened in December 2004. It started with a 9.1 earthquake in the Indian Ocean. This started a tsunami that hit Indonesia, Thailand, India, and Sri Lanka. About 230,000 people died in the disaster.

When tsunami waves start out in the ocean, they may be only about a foot (0.3 m) high. They grow as they race across the water. When they hit land, they can be huge.

QUAKES IN HISTORY

Earthquakes have changed the shape of Earth over the centuries. They've also changed the shape of human society. Many quakes took place before humans could measure their strength.

Around 1500 BCE, an earthquake may have wiped out the capital of Minoan **civilization** in Crete. A quake damaged Pompeii in Italy about 62 CE. Only 17 years later, the city was destroyed by a volcano. A huge quake hit Shaanxi, China, in 1556. It likely killed more than 800,000 people.

FAST FACT

A 1755 earthquake in Lisbon was the start of the modern era of seismology. More people started studying earthquakes and their effects. They also studied how to measure them.

Port Royal in Jamaica was once the biggest city in the Caribbean. It was a base for pirates and the merchants who supported them. The earthquake in 1692 largely destroyed the old city.

NOTABLE EARTHQUAKES BEFORE 1800

1500 BCE: Knossos, Crete
27 BCE: Thebes, Egypt
62 CE: Pompeii, Italy
115: Antioch, now in Turkey
1556: Shaanxi, China
1650: Cuzco, Peru
1692: Port Royal, Jamaica
1693: Sicily, Italy
1755: Lisbon, Portugal
1780: Tabriz, Iran

INTO MODERN TIMES

Scientists began recording earthquakes as the 20th century started. They had new instruments now. In 1903, people created the International Association of Seismology. Scientists around the world banded together. They studied earthquakes. They also created seismographic networks to watch for them.

In 1935, scientists Charles F. Richter and Beno Gutenberg created the Richter scale. This scale was used to measure the strength of earthquakes. By the 1960s, most scientists also came to accept Alfred Wegener's ideas about plate tectonics.

CHARLES F. RICHTER

Scientist Charles F. Richter was born in Ohio in 1900. He moved to Los Angeles, California, when he was young. He studied earthquakes. He also wrote books about them. Although the Richter scale has been replaced, many people still use his name to talk about quake strength.

FIVE EARTHQUAKES OVER 9.0

WHERE	WHEN	MAGNITUDE	DEATHS
Kamchatka, Russia	1952	9.0	10,000 to 15,000
Valdivia and Puerto Montt, Chile	1960	9.5	about 1,600
Prince William Sound, Alaska, United States	1964	9.2	131
Indian Ocean	2004	9.1	about 230,000
Honshu, Japan	2011	9.0	about 20,000

Charles Richter studied and later taught at the California Institute of Technology. He worked at its Seismological **Laboratory**. People still study earthquakes there today.

THE GREAT CHILEAN EARTHQUAKE

The largest earthquake recorded so far took place in May 1960. The quake hit off the coast of southern Chile in South America. It was near the cities of Valdivia and Puerto Montt. Later, scientists estimated the earthquake was about 9.5 in strength.

More than 1,600 people died because of the quake. Most of these deaths happened because of a tsunami soon afterward. More than 2 million people were also left homeless in Chile. The quake caused many millions of dollars in damage.

FAST FACT

More than 100 people also died miles away from Chile. The tsunami hit the Hawaiian Islands, Japan, the Philippines, and more. Waves also damaged the shore in California.

The tsunami that hit Chile after the earthquake had waves as tall as 80 feet (24.4 m). The quake and the waves destroyed buildings and the area's electrical and water systems.

CALI QUAKES

The U.S. state of California has had some powerful quakes over the years. The strongest and most famous California quake so far took place in 1906. On April 18, a 7.9 earthquake hit off the northern California coast.

The quake was caused by the San Andreas fault. Towns along the fault, including San Francisco, had a lot of damage. A huge fire also swept through San Francisco after the quake. It destroyed more than 500 blocks of the city. More than 3,000 people died.

THE WORLD SERIES EARTHQUAKE

On October 17, 1989, the Major League Baseball teams from Oakland and San Francisco, California, were playing in the third game of the World Series in San Francisco. At 5:04 p.m., the ballpark started shaking. Baseball fans around the world watched a 6.9 quake happen live. More than 60 people were killed in the quake. None of them were at the park.

FAST FACT

The San Andreas fault is a major break in Earth's crust in the western United States. It's about 800 miles (1,288 km) long.

About 250,000 people were homeless after the 1906 San Francisco quake and fire.

BEING PREPARED

Do you live in a spot that's at risk for dangerous earthquakes? It's always good to know what to do in an **emergency**. There are also things you can do before an earthquake even happens. That way you'll be ready, just in case.

The American Red Cross has advice for being prepared. This includes putting together an emergency kit and having a household **evacuation** plan. You can also practice finding safe spaces in your home and school.

EMERGENCY KIT

The American Red Cross suggests putting together an emergency kit for natural disasters. Items in the kit could include:

- Bottled water
- Food
- A flashlight
- A radio
- Extra **batteries**
- A first aid kit
- Maps
- Extra money
- Blankets

VERIFIED

The American Red Cross works to help people during emergencies and natural disasters. Its website has much advice for emergency planning. Learn more at **redcross.org**.

The Red Cross advises having an emergency plan for your pets too. Make sure you have pet food in your kit. Know what you'll do to protect your pets during a disaster.

DURING A QUAKE

The best thing to do when an earthquake starts can depend on where you are. If you are inside, stay inside! First, drop down onto your hands and knees. Do it quick, before the quake knocks you down. If you're in a wheelchair, lock your wheels.

Next, cover your head and neck. Cover your whole body if you can. Try to get under a table or desk. (If you can't, get next to an inside wall.) Then, hold on to your **shelter** until the shaking stops.

IF YOU'RE OUTSIDE

If you're outside when a quake starts, stay outside! Move away from buildings and wires. Move away from any gas lines too. Move away from trees or telephone poles. Get down low and stay there until the quake stops.

VERIFIED

The Centers for Disease Control and Prevention (CDC) is a U.S. agency. It works to protect public health and safety. Its website offers natural disaster tips. Visit **cdc.gov/disasters**.

Many people think you should get in a doorway if a quake starts. The truth is that doorways are no safer than other parts of a house.

AFTER THE SHAKING STOPS

After the quake stops, the danger isn't over. Aftershocks can happen minutes to months after a mainshock. If one starts, follow the same advice you did during the mainshock. If you're in an area that can have tsunamis, get away from the water. Go to higher ground.

If you're in a damaged building, get out of the building as soon as you can. Don't go into damaged buildings. If you're stuck somewhere, try to send a text or bang on a wall or pipe.

VERIFIED

The American Red Cross provides first aid training online and in classes. If you know first aid, you can help yourself and others after a disaster. For more information, visit **www.redcross.org/take-a-class/first-aid/first-aid-training**.

First aid is emergency care given to a sick or injured person before other help gets there. It's a good skill to have. Teens can learn first aid too.

UNSEEN DANGERS

Sometimes the dangers right after an earthquake aren't obvious, or easily seen. A building might look undamaged. But if you hear unusual noises, get out! That could mean things are shifting, or moving.

Food and water safety and cleanliness are also important. Food and water can be **contaminated** after earthquakes. If you're not sure food is good, throw it away. Listen to reports about water safety in your area. You may have to boil water before you can use it to drink, cook, or wash.

CARBON MONOXIDE

One unseen danger after an earthquake is carbon monoxide (CO). This is a colorless, odorless gas. It can make people sick or kill them if they breathe in too much of it. Carbon monoxide is produced when people burn certain fuels. This is especially dangerous after an earthquake. Many people burn these fuels when electricity is out.

It's also important to pay attention to your feelings after a disaster. Strong feelings, including those of fear or worry, are normal. Talk to someone you trust if you're having trouble.

LIVING WITH EARTHQUAKES

We don't yet know how to stop earthquakes. Maybe someday, scientists will learn how to do so. Until then, we just need to live with them. While we can't predict earthquakes yet, we can prepare for them. People are learning how to make buildings and bridges that can better withstand quakes. They're studying more about where and why quakes happen. They're working with others to spread good information.

You can do your part to prepare. Make sure your family has an emergency plan. If you can, make sure you have supplies ready just in case. Learn about first aid and safety rules. Remember the drop, cover, and hold advice. Teach it to others too.

Scientists are working on better early-warning systems for earthquakes. New warnings might be possible very soon!

If you like math and science, maybe you could become an earthquake scientist. Take classes in these subjects, especially earth science. Read more about Earth or watch shows about it. You never know what facts might come in handy! Scientists are learning new things about our planet all the time.

GLOSSARY

adjust: To change to fit new conditions.

battery: A device placed inside a machine to provide it with electricity.

civilization: An organized and developed society.

contaminated: Polluted.

continent: One of the seven great masses of land on Earth.

damage: Loss or harm done to a person or property; also, to do harm to a person or thing.

degree: An official title and piece of paper given to someone who's finished a course of study at a university.

emergency: An unexpected and usually dangerous situation that calls for immediate action.

eruption: The act of erupting, or sending out lava, rocks, and ash in a sudden explosion.

evacuation: The act of evacuating, or removing someone from a dangerous place.

laboratory: A room or building in which people do scientific experiments and tests.

landslide: A large mass of earth and rocks that quickly moves down the side of a hill or mountain.

physics: The branch of science that deals with matter and energy, as well as how they interact.

predict: To guess what will happen in the future based on facts or knowledge.

reliable: Able to be trusted; dependable.

resource: Something that can be used.

shelter: A structure that covers or protects people or things.

territory: One of the parts of the United States that is not a state.

volcano: An opening in a planet's surface through which hot, liquid rock sometimes flows.

INDEX